U0392400

Food 80

慢餐

Slow Food

Gunter Pauli

[比]冈特·鲍利　著

[哥伦]凯瑟琳娜·巴赫　绘

李欢欢　牛玲娟　译

上海远東出版社

丛书编委会

主　任：田成川

副主任：何家振　闫世东　林　玉

委　员：李原原　翟致信　靳增江　史国鹏　梁雅丽
任泽林　陈　卫　薛　梅　王　岢　郑循如
彭　勇　王梦雨

特别感谢以下热心人士对童书工作的支持：

匡志强　宋小华　解　东　厉　云　李　婧　庞英元
李　阳　刘　丹　冯家宝　熊彩虹　罗淑怡　旷　婉
杨　荣　刘学振　何圣霖　廖清州　谭燕宁　王　征
李　杰　韦小宏　欧　亮　陈强林　陈　果　寿颖慧
罗　佳　傅　俊　白永喆　戴　虹

目录

慢餐	4
你知道吗?	22
想一想	26
自己动手!	27
学科知识	28
情感智慧	29
艺术	29
思维拓展	30
动手能力	30
故事灵感来自	31

Contents

Slow Food	4
Did You Know?	22
Think about It	26
Do It Yourself!	27
Academic Knowledge	28
Emotional Intelligence	29
The Arts	29
Systems: Making the Connections	30
Capacity to Implement	30
This Fable Is Inspired by	31

兔妈妈有12个嗷嗷待哺的孩子，但家里已经没有食物了。小兔子们饿了，兔妈妈得赶紧采取行动。

“我们去吃快餐吧。”最小的兔子提出建议。

Mother rabbit has a dozen kids to feed and there is no food left. The little ones are hungry and she needs to act quickly.

“Let's get some fast food,” proposes the youngest of the litter.

我们去吃快餐吧

Let's get some fast food

你是说寿司吗？

Are you talking about sushi?

“你是说寿司吗？”兔妈妈问道，“那是唯一我同意你们吃的快餐。”

“妈妈，我们知道你喜欢和食，也知道是日本人发明了快餐。但我们对饭团、紫菜和鱼不感兴趣，我们想吃胡萝卜和各种胡萝卜糖果，而且我们现在就想吃！”

“Are you talking about sushi?” responds his mother. “Because that’s the only fast food I’m willing to let you have.”

“We know you love Japanese food, Mum, and that the Japanese invented fast food. But we are not interested in rice, seaweed and fish. We want carrots and all the goodies that come along with it, and we want it now!”

“从一粒小种子长成一根胡萝卜，你知道需要多长时间吗？”

“不知道，但肯定需要很长的时间，没法很快就让我不饿着肚子的！”

“在凉爽的沙土地里胡萝卜需要生长3个月。如果太热了，它们根本就长不好了。”

“Do you know how long it takes to grow a carrot from a tiny seed?”

“No idea, but certainly too long to still my hunger any time soon!”

“It takes three months in cool, sandy soil. If it is too hot, they will never look right.”

如果太热了，它们根本就长不好了

If it is too hot, they will never look right

为什么胡萝卜有益健康呢？

Why are carrots healthy?

“我喜欢它们鲜艳的橘黄色，吃起来一定味道好极了！”

“无论什么样子的胡萝卜都很好吃。我喜欢胡萝卜，它们是很好的幼儿食物。”

“为什么胡萝卜有益健康呢？”小兔子很好奇。

“I love their bright orange colour; you can see that they must taste great!”

“Carrots always taste good, no matter how they look or what shape they are. I love carrots. They are excellent baby food.”

“Why are carrots healthy?” wonders the little one.

“胡萝卜中的β胡萝卜素含量是其他蔬菜的两倍，而且维生素含量丰富，甜中带点微苦。”

“但是我不喜欢苦的食物。”

“苦味有益健康，这是我从你们外婆那里学到的。”

“They have double the beta-carotene than other veggies, are full of vitamins, and have a sweet taste with a hint of bitterness.”

“But I don’t like bitter food.”

“Bitter means healthy; that is what I learnt from my mother.”

苦味有益健康

Bitter means healthy

每千克大约有100万粒种子

A million seeds in one kilo of seeds

“妈妈，胡萝卜是水果吗？”另一只小兔子问。

“不是，亲爱的。胡萝卜是蔬菜，两年开一次花，每次结出成千上万的小种子，每千克大约有100万粒。”

“哇噢！真的很多。那胡萝卜独自生长吗？”

“Mum, are carrots fruit?” asks another of the little ones.

“No, my dear. They are vegetables and bear flowers every other year, producing thousands of tiny seeds. There are more or less a million seeds in one kilo of seeds.”

“Wow! That’s a lot. Do carrots grow alone?”

“没有人喜欢独自生长。胡萝卜喜欢和香葱、迷迭香、鼠尾草生长在一起。鼠尾草赶跑了所有黏着胡萝卜的苍蝇。”

“因此，尽管胡萝卜需要几个月才能成熟，但只要一撒下种子，很快就能让世界不再饥饿了。”

“是的！甚至不用煮熟，我们就可以吃。”

“No one likes to grow alone. Carrots prefer to grow alongside chives, rosemary, and sage. Sage keeps all the carrot flies away.”

“So, even though carrots need months and months to grow, once you have seeds blowing around, you could quickly feed the world.”

“Yes! And they don't even have to be cooked for us to enjoy them.”

很快就能让世界不再饥饿了

You could quickly feed the world

完全煮熟的胡萝卜最有营养

The most nutritious carrots are boiled whole

"不需要煮熟？现在我们谈论的是真正非常健康的快餐，就像兔笼里给我们提供的大部分水果和蔬菜，连皮都不用削。"

"嗯，很简单。没有切开的完全煮熟的胡萝卜最有营养。"

"煮熟需要消耗更多的精力和耐心。妈妈，我认为我们没有足够的耐心。除非你加些糖，把它做得很甜，不然我们不会等的。"

"No need to cook them? Now we are talking about really super healthy fast food! That's like most of the fruit and vegetables that we are served in our rabbit hutch. You don't even need to peel them."

"Well, it is not that much work. The most nutritious carrots are the ones boiled whole, not cut."

"Boiling means using more energy and needing more patience. I don't think we have that, Mum. Unless you of course add some sugar and make it really sweet. Then we will wait."

“加糖？胡萝卜本身就含有很多糖分了，我们也吃了很多像胡萝卜一样的含糖食物。你们外婆教我用丁香煮胡萝卜，只要你慢慢煮，胡萝卜就会神奇地变得更甜。我喜欢当地的胡萝卜，价格公平。但是我真正喜欢的是新鲜的未加工的胡萝卜，或者慢慢煮熟的、味道香甜的胡萝卜。”

“妈妈，你是说我们应该吃慢餐吗？”小兔问。

“是的，亲爱的，这比快餐健康多了。”

……这仅仅是开始！……

“Add sugar? Carrots are rich in sugar already, and we eat far too much sugary food as it is. Grandma taught me to cook carrots with cloves. That miraculously makes them sweeter, as long as you cook them slowly. I like the carrots that will be available locally, and be sold at a fair price. But what I really like is that we can eat food fresh and raw or cook it slowly for taste and sweetness.”

“Mummy, do you mean we should all rather eat slow food?” asks the little one.

“Yes dear, it is much healthier than fast food!”

... AND IT HAS ONLY JUST BEGUN!...

……这仅仅是开始！……

... AND IT HAS ONLY JUST BEGUN! ...

Did You Know?

你知道吗？

Carrots are root vegetables and even though we associate them with the colour orange, varieties exist that are purple, red, white, and yellow. When carrots were first farmed, they were grown for their aromatic leaves and seeds, rather than for their roots.

胡萝卜是根茎类蔬菜，尽管我们认为胡萝卜都是橘黄色的，但也有紫色、红色、白色和黄色的胡萝卜。人类最初种植胡萝卜是为了收获芳香的叶子和种子，而不是根茎。

Carrots can be stored for a long time, provided they have never been washed. Carrots are 88% water and nearly 5% sugar. Carrots contain beta-carotene, vitamin K, and vitamin B_6.

如果未经水洗，胡萝卜可以存放很长时间。胡萝卜含有 88% 的水分和接近 5% 的糖分，还含有 β 胡萝卜素、维生素 K 和维生素 B_6。

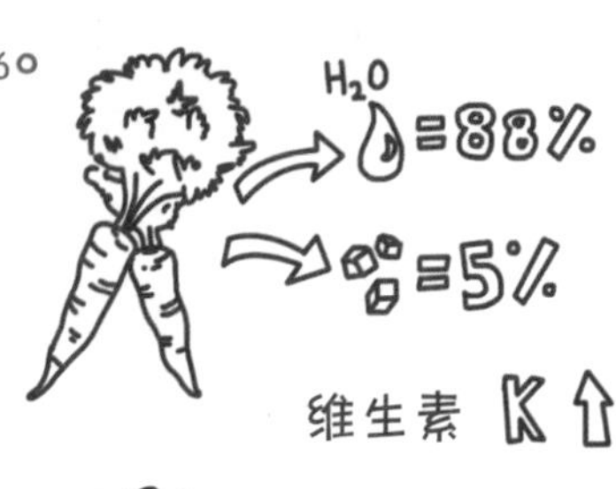

Carrots have been used to sweeten cakes since the Middle Ages. Carrot cake was once voted as the favourite cake in the United Kingdom. Today, carrot cakes are typically made from odd-shaped carrots that supermarkets will not buy.

中世纪以来，胡萝卜就被用来增加蛋糕的甜味。胡萝卜蛋糕曾被选为英国最受欢迎的蛋糕。现在，胡萝卜蛋糕主要由超市不愿收购的奇形怪状的胡萝卜制成。

1 g of carrot seeds contains more than 500 seeds. That is 500,000 seeds in 1 kg. Since every seed can turn into a carrot, it is one of the most prolific vegetables.

1克胡萝卜种子有500多粒，1千克有500 000粒。每粒种子都能长成一根胡萝卜，因此成为产量最多的蔬菜之一。

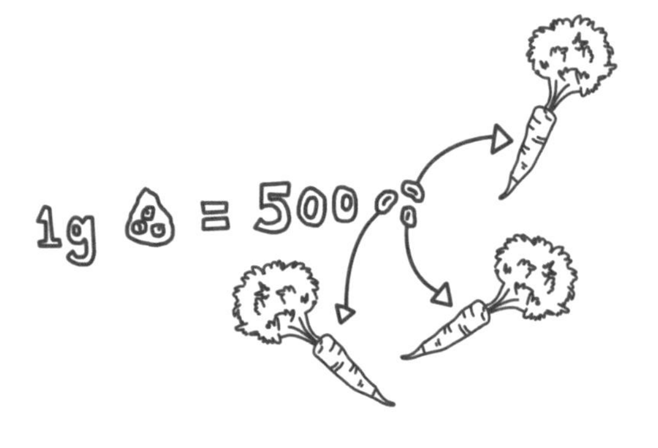

If you give a rabbit a carrot with a green top, it will eat the green and disregard the carrot. If you try to feed a rabbit nothing but carrots, it will die. It is like giving kids only candy to eat.

如果给兔子喂一根带绿叶的胡萝卜，兔子会吃了绿叶而留下胡萝卜。如果只给兔子喂胡萝卜，兔子会死亡，就和只给孩子吃糖一样。

Carrots need to pass through a prolonged cold spell to develop from the juvenile edible carrot root stage to reproductive maturity, during which the plant sprouts flowers. The seeds are embedded in a mericarp filled with oil that inhibits seed germination.

胡萝卜需要经历一段较长的寒冷期才能从幼嫩长至成熟，在这期间会开花。种子嵌在含油分果爿中，受到抑制而无法发芽。

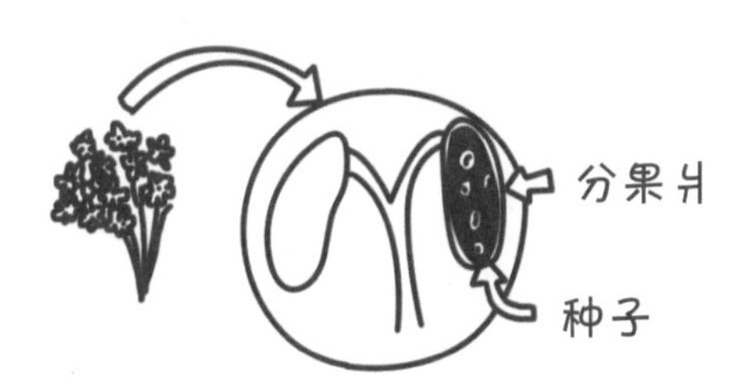

Nearly all carrots you buy in the shop are hybrids and thus sterile. Even if you have the patience to wait for two years to let the flowers grow and harvest the seeds, they are unlikely to grow.

在商店购买的大部分胡萝卜是杂交产品，没有繁殖能力。即使你有耐心等两年，它们也不可能开花结果。

It is easier and requires less energy for the body to digest cooked food. Cooked carrots supply more antioxidants and beta-carotene. The downside is that cooking reduces vitamin levels.

对人体来说，消化熟食更容易，需要的能量更少。熟胡萝卜产生更多的抗氧化物和 β 胡萝卜素，但不足的是维生素含量降低了。

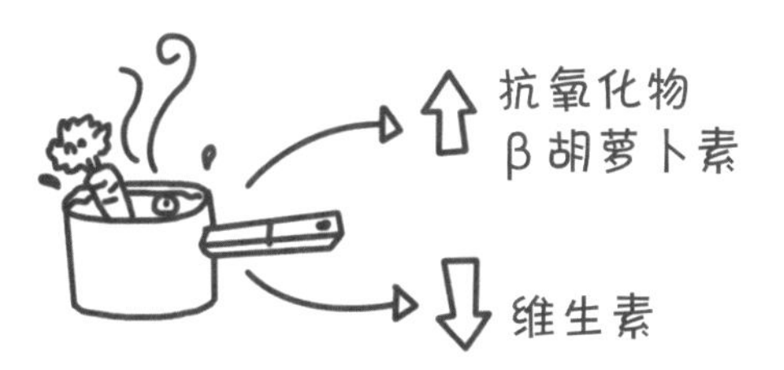

Do you enjoy taking time to have a meal with the family or do you prefer to have fast food and to finish your meal quickly?

你喜欢花时间和家人一起吃饭，还是更喜欢节省时间吃快餐？

马上吃胡萝卜，或在两年时间里用种子种植大量的胡萝卜，三年时间甚至会有更多的胡萝卜。如果必须二选一……你会选择哪个？

If you have to choose between eating a carrot now or to grow plenty of carrots from their seeds within two years, and even more carrots within three years ... what would you choose?

Would you plant carrots alone or would you plant something to grow alongside them?

你是会只种胡萝卜，还是会把胡萝卜和其他植物一起种？

西红柿是水果还是蔬菜？

Is a tomato a fruit or a vegetable?

Do It Yourself!

自己动手！

Let's bake a carrot cake. Ask your mother or grandmother for her favourite recipe or find one on the Internet. To make gluten free carrot cake you can substitute wheat flour with almond and walnut flour. To make your cake extra tasty, decorate it with icing made with coconut cream.

我们烘焙一个胡萝卜蛋糕吧！问问你妈妈或奶奶喜欢的食谱配方，也可以上网找一个。用杏仁粉和核桃粉代替面粉，制作无麸胡萝卜蛋糕。用椰乳味的糖霜点缀蛋糕，让蛋糕更美味。

学科知识

Academic Knowledge

生物学	胡萝卜是有益的伴生植物，开花时会引来黄蜂，消灭许多园林害虫；胡萝卜是二年生植物，发芽需要经历两至三周，第二年才能开出伞状花序并结出种子；杂交种是由两种或两种以上品种人工杂交而成；水果是植物的果实部分，蔬菜是植物的其他部分，如根、叶和茎。
化　学	如果过度授粉，根茎类蔬菜可能含有大量的硝酸钠，危害婴幼儿健康；类胡萝卜素和高铁酸具有抗氧化作用；丁香有生物活性成分能驱赶蚂蚁，并且曾用作牙疼止痛药。
物　理	光谱上，橘黄色处在黄色和红色之间。
工程学	快餐需要严格的物流运输和供应管理。
经济学	单份快餐的利润很低，但成千上万的单份交易提高了利润，这是以规模经济和降低边际成本为基础的：销售量越多，额外成本越低；慢餐运动倡议食用代表本地气候和生物多样性的本地食物。
伦理学	企业说它们会检查硝酸盐含量，但是，检查并没有根除过量的硝酸盐，也没有控制不断过量使用的硝酸盐，而过量的硝酸盐最终会进入地下水；超市拒绝销售某些胡萝卜的原因仅是胡萝卜的外形不好。
历　史	公元前3000年到前2000年，瑞士和德国南部已经耕种胡萝卜；14世纪，中国才开始种植胡萝卜；拿破仑使胡萝卜成为第一种罐装蔬菜。
地　理	荷兰推销橘黄色的胡萝卜，因为橘黄色是荷兰王室奥兰治家族的颜色，是荷兰的国色；慢餐运动始于意大利北部的皮埃蒙特。
数　学	斐波那契的兔子：如果第一个月有一对兔子，12个月后将会繁殖出144对。
生活方式	快餐文化：不花时间准备食物或饮食不合理会导致消化问题；和日本精细烹饪的文化相比，寿司作为快餐而发明；对于什么是水果，什么是蔬菜，厨师和生物学家持有不同的看法。
社会学	我们创作的动物故事并不准确，例如，兔子只吃胡萝卜将无法存活，但动画片让我们相信兔子只吃胡萝卜；橘黄色与欢乐、外向、火焰和危险相关。
心理学	胡萝卜和大棒：奖励和惩罚并用以引导行为。
系统论	喂养孩子不仅仅和食物量有关，也和食物品质、准备过程和食用方式有关；正确饮食能带来更健康的生活。

情感智慧

Emotional Intelligence

小兔子

小兔子们有些迫不及待。他们理解妈妈，知道妈妈喜欢什么。他们也了解自己，知道自己想要什么喜欢什么。关于如何快速得到更多食物，他们想知道更多方法。他们选择食物基于颜色和形状，而健康因素是次要的。他们拒绝食用苦的东西，想要深入了解胡萝卜，兴奋地发现胡萝卜能有那么多种子，能帮助解决世界上的饥饿问题。他们权衡了选择：如果能耐心地等待更长时间，他们能吃到甜甜的东西；但如果想马上吃到东西，食物也许会有点苦味。

兔妈妈

对于养育孩子的方式，兔妈妈很坚定。她知道孩子能吃什么，不能吃什么。她愿意让孩子们思考，并讲清楚了她的决定和标准的来龙去脉。她鼓励孩子们不要光看表面，如食物的颜色。在合适的时间，她说出了孩子们无法想象的惊人事实。兔妈妈分享了代代相传的智慧，用胡萝卜作隐喻，提倡促进多样性和朋友间的友谊。除了加糖，兔妈妈提供了其他选择，希望能给孩子们最好的。

艺术

The Arts

该是画一些疯狂胡萝卜的时间了！不是漫画和动画片里那些无趣的胡萝卜，而是大自然里我们见到的奇形怪状的胡萝卜。它们看似来自另一个世界，但只要你去找一些网络图片和书本插画，这样疯狂的胡萝卜也许真的能在大自然中找到。

思维拓展

Systems: Making the Connections

几千年来，胡萝卜已经成为我们饮食的一部分。蔬菜经销商希望把生产标准化，因此，挑选胡萝卜不仅基于颜色，还看形状。口感好的胡萝卜不被接受，因为它们的形状不标准。无论什么吃法，胡萝卜都是营养丰富的。新鲜的生胡萝卜富含维生素，煮熟后的胡萝卜富含番茄红素。除了健康营养，胡萝卜还有很多价值，提供了控制害虫的方式，最惊人的特点是丰富的种子产量和相对较快的生育期，从种子发芽到成熟，只要两年。胡萝卜可以生吃或煮熟，由于富含5%糖分，不必加糖。胡萝卜有苦的有甜的，就像大多数水果、蔬菜和谷物一样。不管生吃还是煮熟，下咽前的充分咀嚼很重要，适当的胡萝卜准备时间能保证最佳营养价值且易于消化。土地生产什么就吃什么，采用保护地球的方式进行种植，给农民支付合理的工资，这样既有美味的食物，又是我们当地文化的一部分。总之，这样能有更高的生活质量。

动手能力

Capacity to Implement

公元1202年，斐波那契第一次讲述关于兔子的数学，计算出一对兔子一年繁殖了144对，那是很快的繁殖速度。但是现在，让我们比较一下兔子和胡萝卜的繁殖能力。一根胡萝卜结出种子需要两年，搞清楚每次能收获多少胡萝卜。从一对兔子和一根胡萝卜开始，假设在无限的土地上，5年和10年能生产多少吨胡萝卜？多少只兔子？如果我们能合理利用，你认为世界上还会有饥饿问题吗？如果你能算出数量，建立你的论点论据并与其他人分享吧。

故事灵感来自

This Fable Is Inspired by

卡洛·彼得里尼

Carlo Petrini

卡洛·彼得里尼坚信食物应该是优良、干净、公平的，即食物应该有益健康、优质、无污染，以公平价格销售。他以共产主义运动政治活动家身份开始自己的职业生涯。他反对在罗马著名的“西班牙广场”开设第一家速食店，因此迅速成名。他发起了国际慢餐运动，在四大洲经营项目和开展活动。2004 年，在他的出生地意大利皮埃蒙特区的库内格省布拉城，他成立了世界第一家烹饪大学，并被评为 2004 年度《时代周刊》的杰出人物之一。

图书在版编目(CIP)数据

冈特生态童书.第三辑修订版：全36册：汉英对照 /（比）冈特·鲍利著；（哥伦）凯瑟琳娜·巴赫绘；何家振等译.—上海：上海远东出版社，2022
书名原文：Gunter's Fables
ISBN 978-7-5476-1850-9

Ⅰ.①冈… Ⅱ.①冈… ②凯… ③何… Ⅲ.①生态环境-环境保护-儿童读物—汉、英 Ⅳ.①X171.1-49

中国版本图书馆CIP数据核字(2022)第163904号

著作权合同登记号图字09-2022-0637号

策　　划　张　蓉
责任编辑　程云琦
封面设计　魏　来 李　廉

冈特生态童书
慢餐
[比]冈特·鲍利　著
[哥伦]凯瑟琳娜·巴赫　绘
李欢欢　牛玲娟　译

记得要和身边的小朋友分享环保知识哦！
八喜冰淇淋祝你成为环保小使者！